UNDER THE OPEN SKIES

UNDER THE OPEN SKIES

Finding Peace and Health in Nature

MARKUS TORGEBY
and FRIDA TORGEBY

Translation by Christian Gullette

HarperOne
An Imprint of HarperCollinsPublishers

HarperOne

HarperCollins books may be purchased for educational,
business, or sales promotional use. For information, please email the
Special Markets Department at SPsales@harpercollins.com.

Originally published as *Sova ute* in Sweden in 2019 by Offside Press.
First HarperOne edition published in 2020.

Translation by Christian Gullette copyright © 2019.

FIRST EDITION

Library of Congress Cataloging-in-Publication Data

Names: Torgeby, Markus, 1976– author. | Torgeby, Frida, author.
Title: Under the open skies : finding peace and health in nature / Markus
   Torgeby and Frida Torgeby.
Description: San Francisco : HarperOne, 2020
Identifiers: LCCN 2020010269 (print) | LCCN 2020010270 (ebook) | ISBN
   9780063019867 (hardcover) | ISBN 9780063019874 (ebook)
Subjects: LCSH: Outdoor life—Sweden—Jämtlands län. | Forest
   recreation—Sweden—Jämtlands län. | Wilderness
   survival—Sweden—Jämtlands län.
Classification: LCC GV191.48.S93 T67 2020 (print) | LCC GV191.48.S93
   (ebook) | DDC 796.509488/3—dc23
LC record available at https://lccn.loc.gov/2020010269
LC ebook record available at https://lccn.loc.gov/2020010270

20 21 22 23 24   IMAGO   10 9 8 7 6 5 4 3 2 1

"Guard your heart above all else,
for it is the source of life."

Proverbs 4:23

LIFE

INSPIRATION

FOREWORD

I believe in sleeping outdoors, surrounded by tall fir trees, darkness, and cold. Lying on my back and looking up at the stars, watching my breath form thin clouds.

Something happens when you're surrounded by an infinite room. It gives perspective.

When I moved out into the forest almost twenty years ago, I lacked direction in life. I panicked when confronted by all the choices I had to make and all the thoughts I didn't understand. For me, escaping from society wasn't a social experiment: it was pure survival strategy.

Waking up after a long night's sleep on a reindeer skin, my body warm in my sleeping bag, cool winter air on my face, it was like coming home. Everything was clear. Living in the forest felt like something concrete: heat, food, and everything else we take for granted were now my responsibility. It helped me recognize what's really important.

Everything that had been weighing heavily on my mind all those years just disappeared. My blood flowed freely.

During the more than four years I lived under a cotton canvas in the Jämtland forest, I felt a calm that I'd never experienced before. In general, I've never felt as much as I did during that time, never enjoyed it so much.

This is a book about my and my family's way out and what I mean when I say that we lose something if we always sleep in a temperature-controlled room with four walls and a ceiling.

I want everyone to feel the deep rest that I have found in the forest.

Markus Torgeby

CHAPTER 1

THE WAY OUT

It was hard at home and difficult in school.
Life became clearer when I ran across the
rocks on Öckerö, jumped from the bridge
sixty-five feet down into the sea, or steered
my own boat. Nature was my school.

Taking the First Step ⚓ Challenging Yourself

FOR ME, THE JOURNEY started when I was twelve years old and opened my bedroom window. It was just an impulse, but I felt calmer with the winter air on my face. It was as if the cold dampened my worried, racing thoughts and a body that never seemed to stay still. The cold helped me sleep deeply. From that day on, I've always slept with an open window.

Mom had been suffering from MS for a few years and mostly laid on the couch and cried. It happened so fast that she didn't have time to process it. The change from healthy to sick took only a few months. After that, she needed constant care. She had to ask for help with the smallest things. She was sad that life had become something she hadn't imagined, about wanting to but not being able to do things.

Tears ran down her cheeks, her body in slow motion, a slow smile when she saw me. She wanted to be there for me, but she just couldn't. She was so tired.

In the end, I couldn't handle the crying, didn't want to see or hear the tears. I felt a pressure inside that I didn't know how to deal with.

School was like a meadow of knowledge, but when I turned around and looked back, I couldn't see the path I'd walked through the grass. No trace of my steps could be seen; the information didn't want to stick.

There was a bubble in my head where I could do things. But there was also another bubble, and like an hourglass, everything I learned would end up in the bottom glass. I didn't know how to turn the hourglass over, so all the sand remained at the bottom.

Instead, I immersed myself in worlds that I filled with ideas of what I thought was good. Worlds I couldn't get out of. Worlds that created even more pressure and worry.

I stopped eating sweets when I had the idea that I needed to be clean. I avoided eating sugar. I thought I'd get sick, flabby, and lazy. Sometimes I had nightmares that I was so fat, I couldn't get out of bed. I stopped eating potatoes.

I started sailing a dinghy and figured out how it worked: you look at the pennant at the top of the mast if the sail is upright, catch as much wind as possible . . . Simple, clear. I quickly learned how to make my way through the buoys all the way to the goal.

Every win was rewarded with a paper cone of marshmallow candy in different colors. I would come home with a salt-covered life jacket and tired hands, open the white dresser from Ikea, and hide away the cone. I wanted to eat it so badly, but I didn't. By the end of the summer, an entire drawer was full of marshmallow candy, like sweet presents that just wanted to be eaten. As long as I resisted, it worked. I felt strong.

I liked hard rock—Rainbow, Led Zeppelin, Accept—with fast guitars, heavy bass, and strange symbols on the album covers. Then I began to think that the devil was there in the music and that he'd sneak out of the lyrics at night and crawl into my head. He'd gain dominion over me, make me a hollow-eyed addict with needle track marks on my white forearms, and my life would be over.

It came to a point when I could no longer sleep with the hard rock albums in my room. I'd put them out in the hallway when I went to bed. A whole pile of LPs.

Heavy thoughts to walk around with when you're twelve, thoughts I didn't share with anyone and that made me turn inward. I knew that my world would scare those I talked to about it.

I OFTEN WENT to visit my grandma and grandpa after dinner. Their house was next door to ours, so it took ten seconds to get there. Still, I wanted Mom to follow me with her eyes when I darted over there, because I was scared of the dark. I wanted her to watch me through the window the whole way. I never even considered the shortest and fastest way there: through their basement door. That dark basement was more than I could handle.

I knocked on their front door and hoped Grandma would open it as soon as possible. I'd feel a slight panic before I finally saw her through the leaded window pane. Then I'd look back and wave at Mom as Grandma let me in.

ÖCKERÖ

When he retired, Grandfather moored his forty-foot fishing boat *Kristina* in the harbor. I was there almost every day helping, talking, or just hanging around. When I turned twelve, I was allowed to take the *Kristina* out on my own. In the foreground of this image is a shipyard with a crane from which I used to hang by my knees over the water and sometimes over the dry-dock.

My grandma and grandpa lived in my grandmother's parents' home, a three-story house covered with yellow Eternit stucco—Grandpa tore down the old, wooden facade when he grew tired of painting it. In the basement was a laundry room as well as a baking room where they made hard bread and soft wheat bread a few times a year. It was important that the door be kept closed at those times. I don't know how many times I was scolded: "Markus, close the door before the bread collapses and is ruined!" I can picture Grandpa, sleeves rolled up, standing at the oven with a flat baker's peel. Grandma at the milling table, kneading the soft dough. Both red in the face from the heat of the fire.

Grandpa would go to bed at half past eight and Grandma an hour or so later. When I slept over, I always went to bed at the same time as Grandpa. We said our evening prayers and I lay in the middle of the wide bed. He crawled in next to me. He never minded me interfering with his sleep, just as he never minded me coming over unannounced, running up the stairs, and slamming doors.

Grandpa could sleep anywhere: out on the rough sea, in the middle of the kitchen floor after a night of mackerel fishing, or on his back on the sofa without even a blanket. He had perfected his sleeping abilities during his many years on the North Sea.

ON MOM'S BOOKSHELF was a book series about the history of the northern Swedish archipelago.

Black-and-white photos of white-painted wooden fishing boats, all with "GG" for "Gothenburg" stamped on their bows. Facts about the boats: where they were built, how long they were, how much horsepower the engines had.

The Manhood.

Pictures of fishermen mending their trawls and nets, working in teams, shoulder to shoulder. Everyone had their task.

One of the fishery teams was called "Manhood." In the picture, they are standing on a slab of granite in front of a large seine net hung on a wooden stand. All are dressed in cotton pants and thick wool sweaters. Sweaters that their wives and other women knitted during the dark winter months from oily wool yarn sheared from the sheep that grazed the surrounding islands. The sheep were outside in all types of weather, and just like the fishermen, they had been toughened by salty rain and strong winds.

Grandma often sat at the loom in their basement with the radio on. In the kitchen, Grandpa and I could hear the beating of the loom as she worked. The rags she used were torn from thin cotton fabric and stored in large bags on the floor. Some came from white cotton sheets embroidered with AS, after my grandfather's father August Simonsson, who died in 1962. Those sheets had lived a full life on the North Sea, been slept on for a thousand nights, and supported August's head, full of dreams and hopes, and sometimes sweaty from worrying over a bad season. Other rags were light blue and came from sheets that Mom had embroidered with her initials when she was young, before the illness. The rag rugs contained all this. They were woven from memories.

LIFE AT SEA was tough, and often cruel. There was always an uncertainty: maybe I won't make it home again; maybe I'll never see my wife and children again. Some fell overboard and disappeared into the deep, their boots filling up with water. Sinking to the bottom like a soft rock. The sea is a beautiful cemetery.

Finding the right attitude about danger was part of the life of a fisherman. You couldn't let yourself give in to stress or panic when the waves crashed, the boat creaked, and the engine struggled.

I think it's easier to figure out what really matters when you're someone who constantly lives on the edge of disaster. When we live protected from most dangers, we lose perspective—bad internet connectivity or a dented windshield is suddenly a "disaster;" a dead cat becomes a terrible event that requires sick leave.

Grandma told me that alcoholism was a big problem when she was a child. She didn't dare walk the bridge from Öckerö to Hälsö because drunk old men were always hanging around there. In those days, it was normal to drink a lot. Kalvsund is one of ten islands in the Öckerö municipality. It's one of the smallest, and it takes just three minutes to run from one end of the island to the other if you run hard. Despite its tiny size, Kalvsund had eleven pubs.

The free church and the sobriety movement came to the islands at the beginning of the 1900s. The pastors from the Missionary Union and Pentecostal Church preached about a life greater than what we can see, and abstinence as a path to finding God's doorway. Hard liquor was the devil's work and led straight down into fiery hell. Faith and godliness became alternatives to drinking, and salvation swept like a wave across the west coast. My family kneeled before God.

In the northern archipelago of Gothenburg, only a few communities resisted, such as Rö on the island of Hönö. A hundred years later, there are still stories about the residents of Rö. As a kid, I grew up terrified of the people in Rö. The stories made them out to be people with unkempt hair and cigarettes hanging from their lips.

Grandpa would take his fishing boat, the *Klinton*, along the coast of Norway, as did many other fishermen, including those from Rö. The church was not alone in preaching resting on the Lord's Day; the Swedish West Coast Fishermen's Central Union had also decided that the fishermen shouldn't work on Sundays and holidays. But the fishermen from Rö couldn't have cared less about working on Sundays. To them, it didn't matter whether it was the church or someone else who had decided it. They lived by their own laws. Some of the crew of the *Klinton* thought that Grandpa, who was a member of the Central Union, ought to file a complaint that the Rö fishermen were out trawling when they should lie at anchor. To Grandpa's way of thinking, it was none of his business. Whether or not the men fished on Sunday was a matter for them and their conscience.

For most, life became better when the booze disappeared. I grew up two generations later, but in church on Sundays the older generation still reigned with their history and their memories. They were the ones who set the standard, and it was black-and-white and straight and narrow. I understand why now: that at the time, it had been necessary in order to break the negative cycle of alcoholism and addiction.

At the same time, life on the islands had become much easier. The puritanical beliefs and their simple black-and-white answers no longer worked.

For example, I myself never understood why the doorway to hell would open if I played ice hockey on a Sunday.

WE PRAYED FOR MOM, prayed that God would heal her. People from the congregation held meetings at our home where they laid hands

on her body. The atmosphere would be intense with a room full of people joined in prayer, asking for hope and healing.

Every night I sent my own prayer up to heaven: "Lord God, make Mom healthy!" I so fervently wanted to see her well again, dreamed of hiking across the rocks on the west side of Öckerö together and out to sea. That we'd watch the sun set far away on the horizon, the sky red, pink, and orange.

But Mom just spent more and more time on the couch. She couldn't fight gravity. Even the muscles in her throat began to disappear. Yet no matter how small and light she was, she still looked heavy, lying there with her thick brown hair that seemed to stick to the sofa whenever she tried to raise her head.

Sometimes there was a hope in her eyes that she could be healed. I had trouble keeping that hope alive. Doubts crept into my mind, where they took root and grew stronger. In the end, I gave up.

What if it was my lack of faith that made God not want to make her healthy?

WHEN I WAS OUTDOORS, everything was easier. There, I never doubted. There, I never felt inadequate.

From the sea and the wind, I realized that there was something I could never control, only relate to.

Nature was a school I enjoyed, with subjects other than Swedish, math, and religion.

On the schedule was: hang in a tree branch far above the ground or fall down and get hurt; tease an older boy and run away as quickly as possible or get smacked; swim the big waves or drown.

Two hundred meters from our house was a small pond called

AnnaJona. Each spring, we filled it with more water from another pond that was a bit further away. We syphoned off the water with a hose we found at the dump or borrowed from Grandpa. Getting the flow started with my mouth always had the same disgusting taste: first, the aroma of dry rubber hose and then the taste old stagnant pond water with bird feces in it.

In the winters my friends and I skated on the pond. In the summers, it was surrounded by dry grass that we could burn. I remember the feeling of standing there with a lighter in my pocket, about to set it ablaze with my friend Fredrik . . .

We set fire to the grass, and it lit up right away. The grass was dry and tall and burned with twenty-inch-high flames that swept the meadow with thick smoke. Kalle Palm's boat was moored nearby, a dark blue plastic rowboat without a single scratch in the varnish. He hadn't even put it in the water yet; he was planning on painting the bottom.

Suddenly, a gust of wind swept along the ground, feeding the fire until it blazed out of control. Even if Kalle Palm had been understanding about it, we couldn't very well let his boat burn—and Dad would be livid if we did. A sense of flight and adrenaline filled my chest. We were faced with a choice between right and wrong: to take responsibility and try to extinguish the fire, or to escape both the fire and its consequences? Everything was suddenly acutely real. We waved our jean jackets in an attempt to ward off the flames, but the fire just crept closer to the boat. We beat and beat the flames with our jackets, and in the end we managed to extinguish the fire.

Fredrik was covered in soot, and his hair singed. My eyebrows had been burned off, and my bangs were a few inches shorter than normal.

I STARTED RUNNING, and it felt like coming home.

When I ran, everything could be measured: time, distance, finish line. When I ran, everything was easy. No deep thoughts about God and hell, no worrying about my mother's frail body and how long her heart would be able to beat. In the race, it was just me.

The town coach was direct and clear. After only two weeks of official training, I came in fourth at my first Swedish championship. I was happy yet disappointed that I didn't win.

I trained twice a day and constantly strived to improve my performance. When racing, I never doubted. I loved the feeling of my heart beating, blood pumping, my muscles filling with oxygen. Every night I fell asleep excited for the next day's training.

In the mornings I ran a long-distance course across the cliffs on Hönö's and Öckerö's west side, with the sea as my companion and the hard mountain under my feet. Every night, I took the ferry to the mainland to run intervals in the Slottsskogen forest with my training mates, or do long-distance trails in Änggårdsbergen on the other side of the tram tracks, where there were soft paths surrounded by large deciduous trees. From the highest point of Änggårdsbergen I could see the Eriksberg crane, and on the other side, the port of Gothenburg, and beyond it, some way out in the sea, the Öckerö church and the bridge between Hönö and Fotö. The bridge that I and my friends jumped from even before its construction was finished. Sixty-five feet straight down into the water. Butterflies in our chests and stinging feet reminded us of the feeling of hitting the water.

I grew and developed muscles, and I harnessed that new body when I ran. When I stood at the starting line, I knew I was strong.

Everything was up to me. There was nothing to hide behind. It was like stepping out onto a stage with all the lights pointed at me, knowing that what I was about to say was exceptional.

After a few years of hard training, my foot arch collapsed. Suddenly, I couldn't access the physical part of myself anymore. I couldn't run, climb, or swim underwater. I was no longer able to live by what I could see and touch, feel with my pounding heartbeat.

Because of it, I had little resistance when the dark thoughts returned, nothing physical to find shelter in or use as a cushion. My mind fell apart.

⚓

TAKING THE FIRST STEP

Some friends and I turned off the Christmas tree lights of an old man, a guy we knew would get really angry. It was December and dark and cold outside, but no snow. We stood in the darkness a bit away and waited for the old man to come out and start yelling. He never came out. So we headed home, our path lit up by the yellow streetlights.

As we approached the large meadow in the middle of Öckerö, a man with a small dog came walking up behind us. Even as he came closer, we didn't think anything of it. Sixteen feet from us, he sprinted forward and grabbed me by my jacket. It was the man with the Christmas tree lights. "Now I've got you, little devil!"

My friends ran and I panicked. The guy twisted my arm and dragged me back to his house and locked the door behind us. While he called the police, I pissed myself. He called my dad next, who arrived a few minutes later. Dad saw how broken up I was and didn't say much.

Even after showering and eating supper, my heart was still racing. I lay in bed and pulled the blanket over my head, just wanting to hide from those panicked feelings and the memory of the screaming old man. When it got too hot under the blanket, I opened the window. There I was, standing in my underwear in the frigid, winter air. The cold calmed me down, and my pulse slowed. I left the window half open and crawled back into bed.

The room quickly cooled, and I lay under my thick down blanket, my body warm and my head cool. I fell into a deep sleep.

I continued to sleep with the window open and heater turned off year-round. My room was always a few degrees colder than the rest of the house.

———

I believe others can get started on their journey in much the same way. Sleeping outdoors doesn't have to mean going out into the woods or up a mountain. An open window goes a long way towards a restful night's sleep.

Three hours before going to bed, turn off the heater in your bedroom. Open the window, the wider the better: you want as much oxygen ventilation as possible. Turn the lights off. Put a towel or sweater on the floor in front of the door to make sure the cold air doesn't cool down the rest of your home.

Do what you usually do in the evening—or, even better, avoid entertainment. Do something analog instead: take a walk, read a book, knit a sweater, carve a wooden butter knife . . .

Once you go to bed in your cool bedroom, don't turn on the lights. Instead, carry a candle with you. Set the light on the windowsill, blow out the flame, and crawl into the cold bed. Pull the blanket over your head and lie that way until the bed is so warm that you long for fresh air.

Poke your head out from under the blanket and feel the coolness on your face.

———

Sleeping on a balcony, patio, or other flat surface at your home could be the next step.

No special equipment is needed. Just a cold foam mattress and a thick blanket that works down to freezing temperatures. If it's colder than that, a summer sleeping bag with a quilt on top will work. If the place you'll sleep in is a little damp, you'll also need to spread out a small tarp for protection.

—

Check the weather before you decide to try sleeping outside: if it's a summer's evening with an orange sky or a starry winter's night, go ahead. Those are signs that rain and snow are unlikely.

Position the mattress so you have your head against a wall. That will make it easier to fall asleep. In winter, I usually sleep against an east-facing wall, so I wake up with the sun on my face. In the summer I do the opposite, choosing a west-facing wall so as not to be awakened too early by the sun's rays.

Next, get ready for bedtime. Take off your outerwear and put it in a plastic bag at the foot of your sleeping place. Get settled under the covers. Your body is warm, so it will only take a minute or two before the bed warms up.

Look up. The Milky Way, satellites, shooting stars . . . Your frame of reference will change when you're surrounded by an infinite space. The questions you ask will grow bigger.

Lie there until sleep comes.

In the spring, you'll wake to birdsong, no matter where you live. Your blanket will be a little damp, but you yourself will be warm.

Your body will feel relaxed and heavy, your mind at peace.

CHALLENGING YOURSELF

It was a fall evening half a lifetime ago and dark outside. I walked past a lamppost and decided to climb it. I wrapped my legs around the post like scissors and pulled myself up. For each movement, I climbed another foot; the friction between my hands and the metal post was excellent.

Finally, I reached the top and hung there with my feet dangling over the asphalt. My forearms were exhausted. The lamppost rocked. No one saw me. I was alone with my thoughts.

Why was I hanging from a lamppost twenty-five feet above the asphalt? Why was I constantly looking for that kind of resistance?

When I was little, Grandpa climbed up their thirty-six-foot-tall flagpole when the line came loose. He was fifty years old. It was faster to climb up the flagpole than to take it down, he said.

Myself, I think he enjoyed doing what many others couldn't have done, knowing that the flagpole could have cracked in half. There's a charged energy in that kind of uncertainty, in dealing with things that might not go the way you anticipate.

When you do succeed, everything becomes still.

CHAPTER 2

THE QUIET

The darkness scared me. The loneliness
scared me. The thoughts scared me.
But I remained in that fear, and finally
I came out on the other side. I learned
to cope—and live with myself.

I WAS TWENTY-ONE YEARS OLD when I first heard my own heart-beat. It sounded like drumbeats in rushing water.

I had lived in the wild for six months, and outside the hut there was three feet of snow. Everything was nestled under a white blanket, no sharp edges, only soft shapes. The cold stilled the birds, who puffed up their feathers to keep warm.

The forest was quiet.

That kind of profound silence is found only in the dark winter months, and if you want to experience it, you must seek it out.

Nowadays, we've become used to constant noise: traffic, fans, humming fluorescent lights . . . How does it affect us, that lack of real silence? When silence is suddenly scary, even?

How do we live without ever hearing our own heartbeat, the engine that supplies us with oxygen and energy and enables our thoughts? I want to hear a pulse in my ears and feel the rhythm. I want to sense whether or not my heart is racing.

As it says in Proverbs: Guard your heart above all else, for it is the source of life.

WHEN MY FOOT'S ARCH gave out during a training session, my coach completely lost his mind. It wasn't my foot or his training schedule that was the problem: the problem was me and my mind. He could barely contain his frustration.

I never spoke another word with him. A person I saw every day for several years was just gone. No more training.

On Öckerö, I wandered around feeling like there was an entire person perched on my chest, making it impossible to breathe. My tongue swelled and cracked, and it hurt to eat. I didn't have any idea what to do, only that I had to get away from the island, away from everything around me. Away from my mother's illness, my spiritual doubts, and Grandma nagging me to grow up and get a job.

I didn't have any good high school grades to speak of, so I wrote a letter from my heart to the administrator of the recreational leader program at Hålland's community college in Åredalen, Jämtland. I was accepted and moved away from home when things were at their worst—ninety miles north by train from Gothenburg Central Station.

At the community college, no one knew who I was. We were out in the wild, hiking mountains in the fall and skiing during the winter. Slowly but surely, the worst of my frustrations began to disappear.

That year at school, I heard about a man who many years earlier had lived in a tent a little ways from school. That inspired me, and so I decided to move outdoors for real and live alone under a cotton cloth in the forest. I liked the idea of being entirely responsible for myself, where the stakes were high and everything could go to pieces. Sort of like when Grandpa and his brothers fished.

I thought that nature and the seasons would be able to help me find a way to deal with what I couldn't control, what made my breathing labored and my thoughts flutter inside my head.

DURING THE LATE SUMMER, I built a lean-to from fir tree branches and a cotton cloth next to Slagsån, less than an hour's walk from the school. I washed in and fetched my own water from the river. I used riverbed stones to build a fireplace with a smoke hole in the middle of the hut. Then I crafted a bed so I would be a few inches off the cold ground when winter came.

I had saved enough money that I would get by as long as I kept to oatmeal porridge and other basic items I would buy at the grocery store in Järpen. Everything was simple and a matter of necessity: chopping firewood, making a fire, cooking, running, washing, sleeping. I lived without human-made stimuli and stopped consuming radio, TV, and the opinions of others. I was able to be solely in my own head with my own thoughts. I hoped it would help me find my way to my inner self, that core that would be left when everything else was gone.

But even though I wanted this, I soon became restless. On Öckerö, the houses are close, as are the people. The change in lifestyle was intense. It's not that I longed for company, but I nursed a feel-

ing of unsettlement. I walked around talking out loud to myself just to hear a voice. It didn't matter what I did, the restlessness never disappeared.

One autumn day, I walked across a clear-cut on an elevation a short distance from my hut. The pattern on the deforestation machine's tires were still visible on the ground, and thin birches had just begun to reclaim the area. In the middle of the clear-felled field was an old, rotten spruce stump, and I sat down on it. The sun was at its highest in the sky. Just to the east, I could see the white tower of the Undersåker church, and far down in the valley, the dark water of the Indal River.

I sat there, completely still. Silent.

The next day, I went again. After a while, I got the same feeling as I did during math tests in elementary school; my whole body itched. But I decided to stay anyway. Not just for a moment, like the day before, but until it got dark.

My butt, thighs, and back all grew stiff, but still I didn't get up from the stump. It was beyond boring, yet I sat, motionless, not understanding why. Stayed anyway.

Rain or shine, the stump became a daily ritual. Weeks went by. Some days, I sat shrouded in mist, staring into dense fog. Then came the first snow. But I stayed.

For me, moderation has always been a challenge. I'm prone to exaggeration; it gives me a rush that I often seek. Sometimes it's good, sometimes bad.

In this case, I threw myself headfirst into the itching sensation. It was as if the silence and the stump became a powerful lamp that lit up all that I was hiding from and everything that was difficult.

My mom's disease, that I wasn't up to par on the running track, the grief over everything I couldn't control. The feeling of HELL. On the stump, I immersed myself in everything I couldn't handle and let it close.

After three very long months of restlessness, my body felt calm. When it was time to go to the stump, I didn't. I did nothing, but still felt a sense of meaningfulness.

I understood that I would never be completely free, and that there would inevitably be things I couldn't control. But that was okay. And there would always be a room in my head that I didn't want to enter, where there was anxiety, fear, and pettiness. I discovered that was also okay.

Feeling that I was good enough was a strange feeling that I don't think I could have discovered any other way. We're all different, and I'm a slow learner. It took hundreds of hours on a stump for the door inwards to open for me. It's hard to trust that something so important will manifest through being passive. It goes against everything we learn.

For me, that was exactly what it took. The absence of action eventually aligned both body and mind.

WITH THE RESTLESSNESS GONE, only the fear of the dark remained. It was etched into my skin and head, and there was no use trying to run from it or think it away. After darkness fell, it was constantly present, like a neoprene wet suit clinging to my body.

The fear had gradually grown during the fall and worsened the darker the season grew. Before the snow came, it was sometimes so black that I couldn't even see my hand in front of me. I felt as if I was

Life during my four years in the hut was clear and concrete. The branches I used to construct it are still there. A few years ago, a friend found my old jacket there, too. There was an old ATM card in the pocket. In any case, there probably wasn't much left in that account.

being watched by something that wanted to hurt me. Inside my head I saw pictures from horror movies I'd seen as a kid: the murderer Freddy Krueger in *A Nightmare on Elm Street* and the lizards from the TV series *V*.

By the afternoon, I would start to feel the weight and stress of it in my mind. When evening came, it was like the adrenaline rush before a track meet.

I didn't know what to do in order to free myself. With just myself to rely on, surrounded by quiet fir trees and moose moving stealthily through the forest, I felt the fear of the dark like a cold spot a couple of inches above my last vertebra. I fought it, tried to mentally tie it to a flat stone and throw it far across the water's mirrored surface: away, away, away. When that didn't work, I would sometimes go straight out into the darkness, towards the evil that I knew was waiting among the trees. I tried to find this thing that had me in its sights, but I never saw anything, just felt its presence by the creeping sensation on my back. The something that wanted to hurt me.

Sometimes I crawled under a fir tree to calm myself. There, I'd take a deep breath and press my back against the trunk. This helped for a moment, but once I got up again, the fear returned. It went on like that for months.

Finally, I could no longer fight it anymore. I just wasn't strong enough. So, I let the fear into my heart and gave it a hug. I accepted that I was afraid and that things would be the way they were.

One starless evening I slid on my skis and went out into the woods, surrounded by large, dark, silent fir trees. I made it to Lake Norsjön, my body warm from the journey, and headed out on the thick ice.

I sensed the dark steep slopes a bit away, no trees to shield me.

I stood out in the open, surrounded by darkness, and was no longer afraid.

I LIVED IN the forest for over four years. If money ran out, I spent a few weeks as a caregiver at an old age home in Järpen. That would give me enough to live off of for another six months.

I lived physically: I ran across the marshes during the bright season, swam in Lake Helgesjön, and climbed trees. Sometimes in the winter, I skied to the home of an elderly couple I got to know in Duved. It was a twenty-two-mile trip on wide touring skis with a soft skin—over the frozen waters of Helgesjön, up over Björnen, and down to Lake Åresjön. I'd then follow the Indal River home to Ingrid and Bertil, who offered me coffee. Then the same long journey home again.

I chopped wood, fetched water, and used my body to survive. I skied to warm up my system when it was cold. I did everything at a slow pace, which gave me time to twist and turn every thought. Two months could pass before I spoke a word to anyone.

With my head now aligned with my body, I was able to run with a different mindset. I realized I didn't have what it takes to be the best, and that meant that I could run for more than just finish times. When I was out running in the hills or on overgrown forest paths, I wasn't training for any future goals. I didn't long for anything. I was just here and now.

I could never have made that journey if it weren't for nature, solitude, and silence.

But we live in a time when everything inconvenient has to go.

A time that says that if something feels heavy, there's something amiss.

We live our lives high on impressions, constantly on edge as we are fed with new stimuli that have us on our toes. And as long as there's an abundance of stimulation and food, we just continue to stuff ourselves. And we wonder why it doesn't work. Why aren't we happy? I think the answer lies in the fact that we seek happiness when we should be seeking meaning.

If you lose something, you have to look in the place where you lost it, even if it's dark. It won't help to look in the light two hundred yards away.

It is so very typical for us humans to solve what ails us by patching it over and then moving on—or to look to someone else. But it is never another's responsibility that you become the one you are meant to be.

My coach and my pastor had easy and clear answers: train harder, believe more. They set clear-cut rules—follow this training schedule, avoid all sins—and if I followed them all would be good. They never doubted.

Following people with all the answers can feel reassuring in the moment and solve your most immediate problems. For me it did. But none of those people saw the whole picture. If only my coach had realized that life is more than running, had seen me at home with Mom, known how little I ate, known how hard it was for me to sleep at night and that I was failing school, he might have realized that he was pushing me too hard.

I believe that all those who lead others should allow themselves to be wrong sometimes, because there are always things they won't know. It is then up to the rest of us to try not to be overwhelmed by their beliefs. Instead, we should ask ourselves: What might be causing these feelings I have? For the most part, we already have the answer.

For me, the answer was too much of everything: too much worry and too much stimulation. In the end, I was so tired I lost my way. The solution was to reduce, to actively refrain from almost all stimuli. It wasn't easy and it took time, but for some problems, there simply aren't any quick fixes.

We humans have impressions, feelings, and nervous pathways for a reason. It's no coincidence that we feel the way we do. Therefore, we must also have the strength to listen to that which is unpleasant.

Being human includes feeling sad sometimes. All life on earth is cyclical: day is followed by night, after heat comes cold, joy is intertwined with sorrow. That's how it should be. That's how we grow.

If you live so intensely that you need to go on a sunny holiday just to be able to relax, then there's probably something in your everyday life that's the problem.

LISTENING TO YOUR HEARTBEAT

Finding a place in Sweden quiet enough to hear your own heartbeat isn't easy. You either need to go inland, or north, or far out to sea. But there's actually an easier way to hear your own pulse.

—

Fill a bathtub with warm water. Sink down into the water, and take some deep breaths. When exhaling, your lungs should feel completely empty. Then inhale so that your chest is completely filled with air. Hold your breath and lower your head under the surface.

The water will flow into your ear canals and settle as a light resistor against the eardrums. Now you will hear and feel the heart as a clear rhythm; the pulse becomes a vibration through the water.

Your heart is worth listening to. Each heartbeat supplies your body and mind with the oxygen it needs.

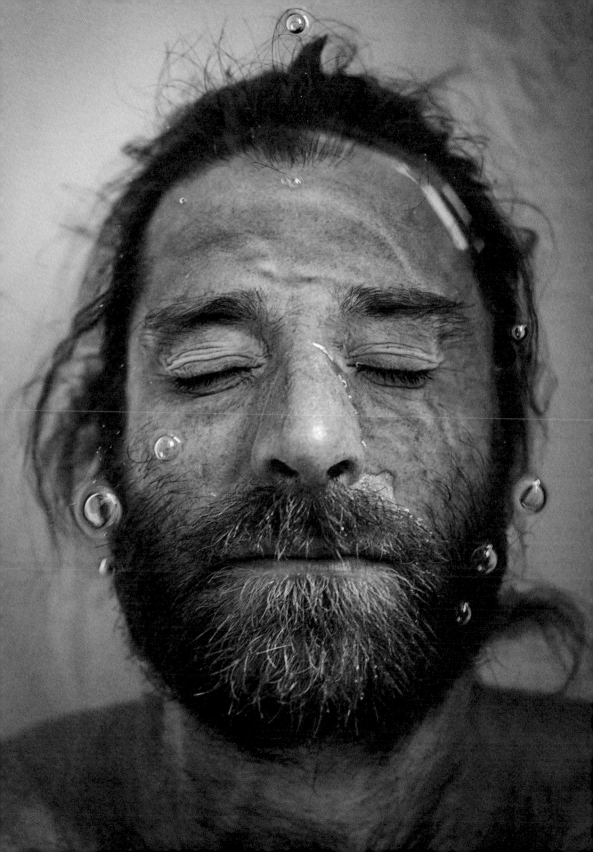

THE SPARK

Only during my years in the forest did I realize how important fire is. That it creates freedom. That a single box of matches contains the possibility of both heat and cooking just about anywhere you are.

I always carried a matchbox or a fire steel when I lived in the hut. Matches are simple to use, cost almost nothing, and can be purchased anywhere. They only have one drawback—they can't get wet. Living in the forest, you quickly learn to store the box in a plastic bag. If I had to choose, I would still choose fire steel.

—

Fire steel is a tool we humans have used since the early part of the Stone Age. It's not sensitive to moisture, has no loose parts that can break, and it never wears out. You can always get a spark, even if your hands are wet and the air is humid. Using fire steel requires only a little training and planning. You can't light the wick on a candle, for example. You need dry kindling against which you hold the steel. To get sparks, drag your knife with a hard, fast pull along the steel, while blowing light gusts of air to help fire get going.

There's a distinct smell when you rub the knife against the steel, like the smell of a braking train. A mixture of steel and fire.

Cigarette lighters, on the other hand, work poorly in the forest.

They are made of components that easily break and can hardly be used when wearing gloves. If it's really cold outside, the gas is too cold, and you'll only get a small, thin blue flame closest to the hole, which then goes out as soon as you try to light something.

BUILDING A FIRE

Being able to make a fire is part of general education. The only thing required is knowledge, matches, love, and a little time. But first, you have to collect combustible material.

The forest is overflowing with things to burn. The rule is to start small.

In Jämtland, where I live, there is a lot of spruce and some birches here and there. First, I go to a birch and peel off fistfuls of bark, which feels a little bit like pulling away loose skin. When I've filled my coat pocket, I have enough to start a fire. Then I break off a few thin, dry fir twigs. They always grow close to the trunk of the tree. You can tell how dry they are from the sound they make when they snap. It doesn't matter if the autumn rain has been pouring down the mountainside—the branches further up protect the thin branches at the bottom. Pick enough to fill a grocery store bag. Then find some fallen branches the same thickness as your fingers, as many as you can carry in your arms. Finally, gather an equally large pile of branches the same thickness as your forearms.

Now it's time to build the fire.

———

If it's windy, find as protected a place as possible. A good rule of thumb is that only a single match is needed to get the fire going.

That's if you've planned correctly. Otherwise, you need to prepare more carefully.

Begin by laying out some strong branches as a foundation. Use your body as a shield against the wind, add the bark, and light the match. When you've achieved a strong flame, carefully add the matchstick-thin spruce branches. Push them down against the fire, and when they catch a healthy flame, add the branches of the next thickness, and then the next. Lay the branches in different positions with some spaces between them, because they need air to burn at their best.

Once the fire is going, you can throw on anything—thin or rough, dry or wet. No need to complicate things.

—

When you've been tending the fire for a while, you might find yourself mesmerized: your gaze will lock on the fire, and your thoughts disappear. It is a wonderful state. It's the same feeling as running when your feet are light, or swimming underwater those first meters when your body is full of oxygen. Time flies by and you feel present. You do not want to be anywhere else.

Fir gives off a lot of sparks, so be sure to wear clothes made of wool or cotton. Synthetic materials are guaranteed to melt, resulting in small fly-like holes.

PORRIDGE

Swedish cross-country skier Gunde Svan might as well throw in the towel; there's no one alive in Sweden who's eaten more porridge than me. During the four years I lived in the tent, I consumed almost 200 pounds of oatmeal per year. Oatmeal is a fantastic food and contains fats, carbohydrates, protein, and minerals. Without oatmeal I would have been dead a long time ago, lying like a crumpled piece of leather under a spruce, dehydrated and forgotten.

It is possible to live a full life for an entire summer consisting on nothing but grains and a little vitamin C in the form of fresh fruits or berries. There's a freedom and airiness in that idea, that you don't need much in order to have a good life. During periods when I had some money, I counted everything in oats. A pair of new sneakers for 1,000 Swedish crowns ($103) or 220 pounds of oatmeal for the same price? Thinking this way can sometimes help us see what's really important in life.

When I lived in the woods, I used an old twelve-cup steel saucepan, blackened with soot. In the beginning, it had a Bakelite handle, but it disappeared rather quickly. It melted in the fire, and the whole tent stunk of plastic for a couple of days. Instead, I used a woolen mitt when I lifted the pot out of the fire. Today, the pot is worn-out and dented, but it has a patina and holds a lot of memories.

—

Most pots made for outdoor use are made of aluminum with a coating of Teflon, to make them easy to carry and wash. They work well with gas kitchens and plastic cutlery, though not with open fire and a steel spoon. That destroys the Teflon, and it comes off in microscopic bits that will end up in the porridge. Although steel pots are heavier, they work best if you're cooking porridge over an open fire. Steel never breaks, and you can easily remove stuck-on porridge with sand or small stones.

—

I've tried most grains at least once: rye, whole buckwheat, crushed buckwheat, and bran oats that I ground in a coffee grinder. Rye requires a longer cooking time. Buckwheat really needs to be soaked overnight and easily boils over and extinguishes the fire when cooking. In short, I prefer regular oatmeal: it cooks the fastest, grows in Sweden, and feels most satisfying in my stomach. Gyllenhammar Oatmeal is the most expensive and has the thinnest and smallest grains, which makes them quicker to cook and makes a nice and sticky porridge. The cheaper varieties have larger grains and need a little more time over the fire. Often, it's the thickness of my wallet that determines the brands I've used.

—

Also when it comes to porridge water, I've tried most things: stream water, lake water, muddy water, water from melted snow or icicles, even salt water. Salt water was a miss; you can probably only use that to cook potatoes. In Sweden, most lake and stream water is drinkable, but in the summer the sun warms the water and the muddy taste increases the warmer it gets. Porridge that tastes like pike isn't that great—a little cinnamon, however, works fine to hide the earthy taste. If you don't have access to running water in wintertime, melting snow works just fine. Just keep in mind that melted snow has a very bland taste; just a few grains of salt will make a huge difference.

——

In the beginning, I went with a classic porridge of water, salt, and oatmeal, but over time I started to work with different spices like cinnamon, cardamom, oregano, and thyme, with and without salt. Eventually I discovered that the porridge tasted better without salt, and that cinnamon and cardamom worked best.

I also tried our Swedish fruits, as well as some imported ones: apples, pears (they get too mushy), plums, blueberries, lingonberries, cloudberries, bananas, figs, and raisins. I also tested some vegetables, mostly because they were cheap: carrots, parsnips, and potatoes. They weren't nearly as good, but carrots at least give a sweetness to the porridge if you add them early on.

——

In my opinion, you'll achieve the ultimate porridge experience by following these steps:

You'll need a steel pot, a spoon, a knife, milk, oatmeal, cinnamon, raisins, some figs, and a hard apple. Find a stream with cold running water. Make a fire that smokes a little—the smoke flavor adds extra spice to the porridge.

Pour two cups of water into the saucepan, and put in some cinnamon at the same time, along with a bunch of raisins and some figs

that you've sliced in half with the knife. Let it boil until the water has taken on the color of the dried fruits, then toss in the diced apple. Let it cook for a minute. It can be difficult to control the heat of the fire, so you may need to pour in a little more water. Add one and a quarter cups of oatmeal, and cook until the porridge is firm. Pour in milk and eat directly from the saucepan until your stomach is full.

—

Supplemental Lesson: add a few branches on the fire just before the porridge is finished. Spruce gives off sparks, and if one lands in your porridge, you've got it made. If you're outdoors in late summer, you should of course add in blueberries, raspberries, or cloudberries. In the winter, you need more energy, so you can top the milk with a little cream.

DRYING CLOTHES

When you're in the woods or just out in nature—and it's not summer and clear weather—everything you have on your body becomes damp. If you don't do something about it, your socks, clothes, and shoes will sooner or later end up wet.

During the warmer parts of the year, the worst thing that will happen is that you'll feel uncomfortable. However, during wintertime, wet clothes can mean the difference between life and death. It may sound dramatic, but keep in mind: you can sit in a 210-degree (Fahrenheit) sauna without a problem, but in the same temperature water, you'll die. Since water is just as good at conducting cold, it's even more important to stay dry the colder it gets.

There are a few different techniques for drying your things while outdoors. I've tested most of them, and this is what I came up with.

—

First and foremost: there's a difference between being damp and wet. Never be wet, at least not unnecessarily.

Use your head. It doesn't matter if you're going for a hike or walk, skiing, or chopping wood—you'll generate moisture. But unlike rain, snow, and fog, you can regulate your own body's moisture yourself. If it's cold and you're going to be active, wear as little clothing as possible.

BARBECUED

Drying your clothes cowboy style, with socks draped over a branch or hung by a string over the fire, is a bad method.

Better to have socks on your hands and sit close to the fire.

Then they are heated up both from the inside and out—and won't go up in flames.

If I go skiing with a sled and it's windy, I usually wear only two thin wool sweaters plus a cap and mittens. If it gets windy, I take off a wool sweater and put on a cotton jacket instead. If I get too cold, I put something back on. If I start to sweat, I take off my hat. I constantly regulate body temperature by adjusting clothing and effort level.

My back always gets a little sweaty under my backpack, and as soon as I stop to take a break, it gets even more sweaty. So, I immediately remove my upper body clothes and stand bare-chested for a minute until the sweat evaporates. Before I get dressed again, I shake the sweaters to remove some of the moisture. I do whatever it takes to stay as dry as possible.

—

If it rains, I prefer a long rain jacket to a rain suit. If you wrap the entire body, you can get wet from the inside instead. Instead, I use a rain jacket that goes down over my knees, which allows for air to enter and circulate, evaporating the moisture.

If I'm out on a trip—and it doesn't matter what time of year it is—I always bring a dry change of clothes in a waterproof bag in my backpack. This includes a shirt, a thicker sweater, and a pair of socks. That bag is sacred and must not be opened before the tent or shelter is erected. Only when I know I'm protected from precipitation and will not generate more sweat will I put on my dry clothes. Then it's also time to dry the wet ones.

—

Drying socks is quite easy and satisfying. Put them over your hands, and hold your fists up towards the burner or fire while you're warming up your food. That way you'll be able to feel if you're heating them too much and avoid ruining your socks. It works just as well to drape them over a branch or hang them on a string over the fire.

My favorite method is to use a heat-resistant water bottle that I fill with boiling water and then pull the wet socks over the bottle. The colder it is, the faster they dry, it takes a maximum of five minutes. When the socks are dry, I make tea using the water or use the bottle as a hot water bottle at the bottom of my sleeping bag.

Clothes take longer to dry. The best way to dry them is to light a fire and sit close to the flames with the wet clothes on your body. The heat from the fire and your body meet, and the moisture evaporates. This way you'll also feel if it gets too hot for the clothes. If you must hang your clothes on a string near the fire, be sure that the fire doesn't damage them.

—

The next morning you put your evening change back in your backpack again. It doesn't matter if you've managed to get your day clothes completely dry—you can still wear them. The sacred bag is your safety: whatever happens, there will be dry clothes when the evening comes.

Planning ahead is the basic requirement regardless of the season.

WOOL

Wool. A material that we humans have used for thousands of years. Cool when it's warm outside, and warm when it's cool. It keeps you warm even when damp and can be used for anything from underwear and sweaters to socks and hats.

Sheep eat grass and help keep the landscape manicured. We shear their hair, card it into wool, and make clothes. This goes on for years and years until it's time for slaughter and we eat them. Grass, clothing, meat—an optimal cycle.

I love wool. I never use anything else closest to my body. Why would anyone? I can't see a reason.

—

Wool can bind up to 30 percent water and still feel warm. No other material can compete with it. Wool also contains the substance lanolin, which is like having your own soap when it comes in contact with moisture, as it rejects dirt. If you have a good quality wool garment, it will wash itself—all you have to do is air it for a while. Wool never smells bad.

When I feel wool against my skin, I think of the fishery team "Manhood." That hundreds of generations before me have worn wool testifies to its unrivaled nature.

—

Wool base layers are often more expensive than synthetic ones, but you get what you pay for. Manufacturers use all kinds of technical terms to describe the latest synthetic underwear, but it doesn't change the fact that compared to wool, it's crap.

—

Despite the fact that wool almost never needs to be washed, there's a perception that it's too difficult to hand-wash woolen garments or run them separately in the washing machine's delicate cycle. Therefore, wool is sometimes treated with a plastic film around each fiber, so it can be machine washed at 100 or even 140 degrees. That's just as ridiculous as if we humans were to clothe ourselves in cling wrap because it would make showering off dirt easier.

Beware of treated wool. It's useless.

Often reading the laundry instruction tag will be enough, but if you want to be absolutely certain, pull off some wool fibers from the garment and light them with a match. If the fibers self-extinguish and then crumble when you rub them between your fingers, the wool is as it should be.

FOOTWEAR

My whole life I've had cold feet. It's as if the blood vessels in my feet are somehow more exposed than they are on other people, and absorb the cold when it's below freezing.

The first winter in the hut, it didn't matter how much I moved or how sweaty my upper body was; my feet still hurt. Every day, I had to light a fire and crawl into the sleeping bag to warm my chalky-white toes.

Finally, I started to wonder what I was doing wrong. There are peoples who've lived for millennia in even colder places, such as the Sami and Canada's indigenous peoples. How do they manage?

One day I came across a Sami I know at a grocery store in Järpen. I asked if he had any thoughts on how I could keep my toes warm. What shoes should I buy? He didn't say much about winter boots, but instead he talked about winter clothes. It was then that I realized that rather than wearing a winter boot that was stiff and hard like a wooden plank, I should be wearing a boot that was soft and pliable, like a thick, wool sweater but for my feet.

At the library in Järpen I read books about Sami and Finnish loggers. The loggers used shoes made from felted wool that they shaped after their own feet, like walking in really thick socks. The Sami used soft hide shoes made of leather from reindeer legs and insulated with hay. I got a tip that a man in Kiruna named Isak Salming custom-made them, so I bought a pair.

Walking in such shoes is like walking barefoot. The foot can move in all directions and feel the snow and ground through the soles. When the next winter arrived, my feet no longer felt frozen.

———

The winter boots sold today are usually rigid, designed not to twist, and banana shaped for us to roll forward onto the ball of the foot. Exactly the things that lead to cold feet. If you're having the same problem, try to find soft boots made from a material that breathes well. The sole should be pliable and wide, so every muscle in your foot has room to move. Your feet will be warm.

———

In the summer, I usually walk barefoot, or in sandals if the surface is hard. Rubber boots with the shafts folded down if it's wet or cool.

Many of today's outdoor boots are too rigid. This works if you're going to hike in southern Europe where it's rockier. But since Swedish terrain is often soft and a little wet, it's better to have pliable footwear that frees all the muscles in the foot. A foot should be trained and become strong—that's its job. The idea that boots can provide stability for your feet is just a sales pitch.

———

If you have problems with your feet when carrying a heavy backpack, then getting insoles or new boots isn't going to help much. The pain is there for a reason: listen to the pain and try to figure out the cause. Most often, the explanation is that your feet are too weak. Walking barefoot will make them stronger.

If you haven't walked barefoot outdoors in a couple of years, it will immediately bring back childhood memories. It never fails. Your feet remember how the grass felt.

SLEEPING BAGS

If you go into a sporting goods or outdoors shop, there are countless sleeping bags to choose from. Synthetic bags and down bags of varying sizes, lengths, thicknesses, and shapes, with various technical solutions. All are optimized for different temperatures, and if you believe the manufacturers, it's as if each occasion and season requires its own sleeping bag.

———

During the years I lived in the hut, I used a down sleeping bag I found on sale for SEK 1,500 ($103). The most important thing for me was that it could keep me warm in the winter, so I opted for this German-made winter sleeping bag that could handle up to minus 31 degrees. The outer material was a thin synthetic fabric that felt like silk, light blue on the outside and on the inside, the same deep purple as the capes worn by the soldiers in Astrid Lindgren's *The Brothers Lionheart*. The bag was actually made for tall people, but I, who am not very tall, liked the thought of having the extra space by my feet for drying sweaters and socks during the night.

During the warm and bright summer, when the birds sang in the forest through the night and the sun rose early every morning, I used the winter sleeping bag as a quilt. Except for wool underwear, I slept naked on a reindeer skin.

When the birds started singing less and preparing for winter, I zipped up the sleeping bag and put on a thin wool sweater and thin socks. Autumn is an easy time to sleep outdoors. It's dark, still, and cool. The perfect conditions for deep sleep.

When autumn turned to winter, it took a few weeks before the cold settled in, but once the snow had fallen like a silent, white blanket between the trees and the temperature crept below 5 degrees, I would pull on my wool underclothes and a thick cap on my head. On the days when it was really cold, when the spruces would make a sound as if they were cracking, and my piss froze to ice and my beard stuck to my pillow, I wore two hats and extra thick socks.

———

The trick to not freezing when it is so cold is to try to not circulate the warm air closest to your body, and lie completely still. To just relax and let the warm blood circulate out to your toes and fingers. Take long, slow breaths through the nose, which is like a heater that transforms cold air to warm before it reaches the lungs. When it's minus 18 degrees or colder, the top of your nose may get a little stiff, but this usually works as long as you take slow breaths.

In the spring, when the days were warm but the nights still cold, I felt an intense pressure in my body after the winter cold. I couldn't get cold. Even though it was below freezing, I lay with my sleeping bag down at my navel and just one, thin wool sweater to cover my upper body.

You will, in other words, do fine with a winter sleeping bag. The rest you regulate yourself by choosing the right sleeping clothes and opening the bag differently at different times of the year.

A good sleeping bag is simple, and the only thing that can really tear is the zipper—so be careful with it. There are sleeping bags without a zipper, and if it weren't for the fact that you can't use them as a blanket, they would be the best bargain.

Synthetic or down, it doesn't really matter. I prefer down, which is light, soft to sleep in, and despite its volume, can be scrunched down into a small ball. The disadvantage is that down collapses and loses all its benefits when wet.

A synthetic sleeping bag is constructed differently and never collapses—a big advantage. The downside is that it's heavier and more compact, leading to a less restful sleep.

—

One problem when I lived outdoors year-round was the moisture that spread to my down bag. It only occurred in winter as cold air doesn't bind moisture.

The body gives off heat, and when it meets the cold outside the bag, condensation forms in the same way that a glass bottle with cold liquid beads moisture when in the sun. This meeting of hot and cold air occurs in the sleeping bag insulation. Day after day, body heat encounters the cold outside and the insulation becomes more and more moist and humid, and after a few weeks the down collapses. It feels a bit like sleeping inside a popsicle. Not a pleasant sensation.

Of course, most people don't sleep outside in minus 36 degree temperatures week after week. But for those who may be wondering, I solved this problem the next year by buying a thin, synthetic bag. I pulled it over the down sleeping bag in the coldest weather.

The condensation settled on the summer sleeping bag instead, and the down one was unaffected all winter.

———

There are two more things to keep in mind when it comes to sleeping bags.

If you wash them too often, they lose their effectiveness. There-fore, treat your sleeping bag the same way you would a regular quilt, which you protect against dirt with a duvet cover. So sleep inside a duvet cover inside the sleeping bag if you plan to use it a lot. A du-vet cover is much easier to wash and dry than a whole sleeping bag.

If you plan to light a fire near your sleeping bag, you need to pro-tect it from burn holes. Use a wool blanket as a quilt to protect the bag from sparks. A cotton blanket also works.

SLEEPING MATS

I've slept on most types of sleeping mats, from spruce beds to air mattresses. When I was a little boy, I was in the Scouts, and each summer at camp somewhere in Bohuslän, I used a regular half-inch-thick rolled foam mat. It's the world's most commonly used sleeping mat for several reasons: it's cheap and easy, takes up little space, can't get wet, and lasts a long time. The drawback is that plastic foam pads are tough to sleep on—after just one night, your body will feel as stiff as if you'd slept on a rag rug on a kitchen floor.

Nowadays, there are inflatable models that are like a mixture between an air mattress and plastic foam pads. They are both more expensive and more comfortable to lie on, but the downside is that they're a little sensitive: if it gets a hole, it's like sleeping on a plastic bag. Sure, there are repair kits, but I think your bedding should be more robust. When you're tired and ready to crawl into the sleeping bag, the last thing you want to feel under you is sticky adhesive and rubber patches.

—

If you know you're going to stay in the same place for an extended period of time, I would recommend spruce. It takes a little time to make a sleeping bunk out of spruce branches, but done right, it feels terrific. To lie on an eight-inch-thick bed, surrounded by the

pungent smell of spruce and resin, is like being hugged by a tree. The trick is to look for branches that aren't too thick and extend enough to provide an intermediate layer under your upper body so that you have a soft bed that isn't too uncomfortable.

But my absolute favorite foundation is untreated reindeer skin. Reindeer is organic and always emits a pleasant warmth.

—

Reindeer live year-round in the forest and in the mountains. They can run in deep snow across the plains with their large hooves, they're resistant to mosquitoes, and can survive minus 49 degree temperatures. This resilient animal survives on lichen and snow and sometimes wanders out onto the road to lick salt.

The reindeer can stay warm because each hair traps air, which is why their hides are so comfortable to sleep on. The hairs aren't lean and shiny like cat or human hair, but a zigzag shape, like a thin type of sea reeds.

I slept on reindeer skins every night during the four years I lived in the forest: a reindeer bed instead of Hästens. My last spring in the forest, I woke up every morning with little hairs all over me; they fell off in the thousands. The skins became thin and stopped warming me as well. In the end, it was like lying on a piece of cardboard that smelled of dried ham. Then it was time to get new ones.

—

I buy my reindeer skins from Per-Erik in Järpen, a tough-as-nails Sami man from Tossåsen's Sami village with a shaved head and kind, brown eyes. The skins should be untreated—that way they're more resistant to moisture, and the hair lasts longer. I stretch out my hides with the skin side out and nail them to the house wall, high enough so the foxes and other four-legged animals can't get them, and then let the air and the autumn sun do their thing. During the winter, the birds peck away any last bits of fat; they cling to the skin with their sharp little claws and eat themselves fat for the winter. It takes a few weeks before the skins are ready, when they're dry and stiff and still have a faint scent of animal to them.

The Sami people slaughter reindeer from September to October, the best time to get hold of fresh skins to dry yourself. Otherwise, you can just drive north and buy ready-made ones at any gas station north of Sveg.

—

Reindeer skins are a bit shorter and heavier than other forms of bedding, and are sensitive to rain. But for me, it doesn't matter. The warmth and the feeling they give is superior; I wouldn't want to sleep on anything else. Two skins are sufficient for your whole body.

Keep in mind that synthetic base layers and reindeer skins make for a bad combination. Your base becomes static, the hair is pulled to the material like a magnet, and you turn into a hair ball. Wool underwear and reindeer skins, on the other hand, are like a harmonious and glitch-free marriage.

THE AX

If I had to settle for only one tool, I would choose the ax. It can be used for everything: cutting, chopping, splitting, slicing.

With the help of a single ax, you can build a house that stands for five hundred years. You can chop the wood that allows you to cook and keep warm. In the forest you can survive for months with only an ax, some matches, and forty-four pounds of oatmeal.

I find it a dizzying thought that a piece of steel mounted on a wooden shaft holds that potential.

—

For an ax to do its job, it must be sharp. All that could slip poses a danger. The edge is sufficiently sharp if you can use it to cut off a little of your fingernail. To maintain it, you need a whetstone, as well as a file to remove notches.

—

When working with the ax, you should have it close to your body. That way, you maintain control and don't hurt yourself. Spend several hours getting to know your ax: a good exercise is to try to hit the same mark on the chopping block time after time. Try to hit the mark even after you move a few inches or go off to drink water. Look directly at the spot you want the ax to meet.

If you've cut down a tree and want to remove the branches, stand on one side of the tree and chop the other side, with the trunk protecting your legs. Don't stand like a curved cheese puff when you chop, but try to engage your abdominal muscles. Think of your torso as a cylinder that needs an internal pressure to stay strong. That way you protect your back.

When I chop wood without a chopping block, I always sit on my knees with a log laid out horizontally on the ground in front of me. If it's cold, I rest my knees on my gloves to protect the joints from the cold. With one hand I hold the piece I want to split against the log, and with the other I chop. This is a foolproof way to split firewood: you can't cut yourself in the leg and don't risk planting the ax in the ground. Perfect for when you're chopping firewood into smaller pieces that are easier to light.

—

When buying an ax, choose a medium-size one with an eighteen- to twenty-inch-long wooden handle and a five- or six-inch-wide steel blade. A handy ax is easy to pack in your backpack and can still be used for most things: chopping down small trees, cutting off branches, and splitting firewood up to six inches in diameter.

You'll be able to use a handmade ax with a wooden handle for your whole life and then pass it on to your children when you are old. In turn, they can pass it on to theirs. An old ax holds many memories, so treat it with respect.

**REPLACING
THE HANDLE**
Buy a new handle at a hardware store or a thrift shop, or make one yourself. Use strong materials like oak or hickory. Make a notch on the top of the handle with a saw, roughly an inch long.

DRIVE IN THE HANDLE
Bang the grip of the handle into a stump to make the blade come down as far as possible.

USE A WEDGE

Drive it in as far as you
can. The wedge should
be made out of a hard
wood as well.

SAW OFF
THE WEDGE

Saw along the blade
of the ax. If the blade
begins to feel loose after
a while, immerse the ax
in water overnight, and
it will be as good as
new again.

KNIVES

During my childhood, I saw my grandfather's folding knife every day. He kept it in his right pocket so that it was always at his hand's reach. A splinter in your finger? Out came the knife. A branch needs trimming in the garden? Out came the knife. Straggly eyebrows? Out came the knife as Grandma shook her head, saying, "Rune, what are you doing?" Sunday pants too long? Out came the knife, making Grandma really upset.

Grandpa used the folding knife as cutlery when he ate. It came out when he peeled potatoes, when he carved my first slingshot, and when he mended with yarn. There he sat, down in the basement, in a green apron over his shirt, sleeves rolled up, and with the sea report at maximum volume: Norra Kvarken, Bottenviken, Gotska Sandön. Thin nylon rope and fifty-five yards of yarn in a pile on the floor.

I think Grandpa loved his knife. It was like an extension of his rough hands. If I ever asked if he had his knife on him, he'd shake his head and mutter: "A fisherman without a knife?"

He even had it in his dress pants for the Sunday meetings, in case there was something you'd need to cut down. He'd just have to reach into his pocket and out came the black folding knife ready for anything. It was always sharp, but nothing to be afraid of.

———

Personally, I'm more attached to my ax. An ax is something I can't do without, whereas a knife is just good to have. I have a knife for cutting cords, chipping wood, or lighting fire steel. Above all, it's mostly just a treat. A knife can transform a piece of wood into anything from a porridge ladle to a child's toy.

—

Carving wood requires real attention. Always keep the knife close to your body when carving: it gives you more control. Carve with bent arms, and use your chest muscles if you need to cut away something particularly hard.

When choosing a knife, think less is more. It's easier to control something that is handy and closer to your hand. If you don't live in a rain forest, big adventure knives are just awkward and get in the way.

Look for a soft wooden handle that is comfortable in your fist, preferably with a stop between the handle and the knife blade so you don't slip out over the edge if you lose your grip. Or buy a regular plastic-handle MoraKniv, which costs less than a hundred kronor ($11). They're found almost everywhere for a reason: they can be used for almost anything.

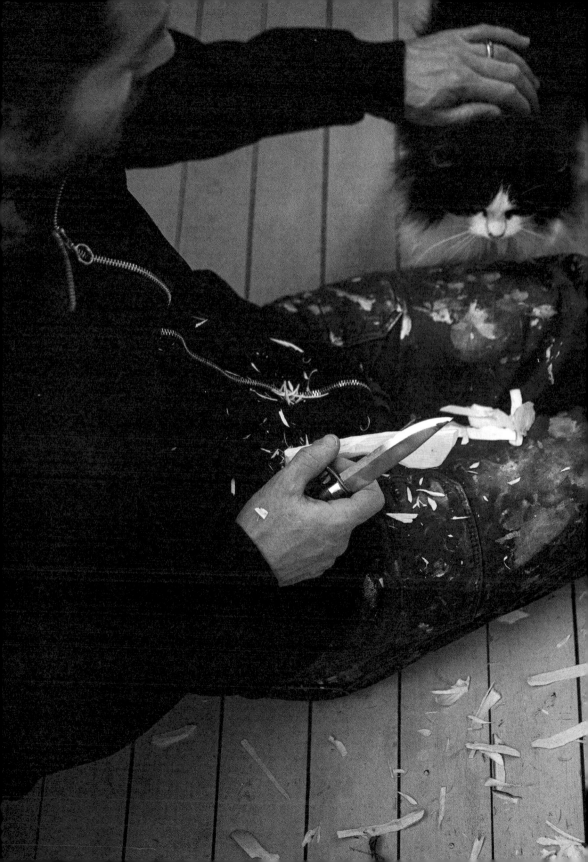

CHAPTER 3

TWO WORLDS

I have become *we*. Porridge and spruce replaced with a refrigerator and underfloor heating. We exist in both worlds now, without walls. City and country. Parents and children. Water from the tap and water in Lake Helgesjön. The bed in the house and reindeer hide in the forest.

Climbing Trees ❭ Tapping Sap ❭ Edible Plants
Doing Your Business Outdoors ❭ Sleeping Outdoors in the City
The Outdoor Bed ❭ Drying Firewood ❭ The Stump

NOW, I'M FORTY-TWO YEARS OLD, in the middle of my life. I don't see any shadows behind me yet, but no doubt Death has already visited and taken my measurements for my burial suit. My hair has started to gray at the temples, and I need to trim my eyebrows more often.

My family and I live in a rebuilt mountain village up in Jämtland, a few kilometers from the place where I once lived in the hut. We have electricity but use fires for heating.

When I get up, it's still a bit chilly in the house. I place fistfuls of birch bark in the stove and add in coarser wood as the fire burns stronger. I can smell the smoke before I close the door. When the house goes from cool to warm and I sit near the stove and feel the heat on my chest—that's the highlight of the day.

I wake the children. Signe, who's twelve and almost a teenager, gets up right away. Her middle sister, Helga, ten, stays in bed and has a hard time getting up. Birgitta, the youngest, gets to sleep for a while longer.

A moment later, the big sisters go down the hill to meet their ride to school, backpacks on their backs. The same morning routine for the past six years. The children grow and the time passes.

I pull on a pair of boots and a thick sweater; it's time to let the chickens out. The rooster has a bright red comb on his head and likes to flap and puff up his wings with confidence. He keeps a close eye on his chickens, but still the fox gets hold of one of them from time to time, and the only thing left is a small crater-shaped pile of white feathers.

Our cat, Björk, has long black fur and a small, thin body, a white stomach and paws. One day, she disappears. The children are heart-broken: Could the fox have taken her? We call for her in the dark before we go to bed. No answer. The days pass and she doesn't come back.

A week later, Signe says she heard something from inside the forest a few hundred yards from the house. I run there, hearing a faint mewing, but I can't tell where the sound is coming from. I walk between the trees and finally realize that the wailing is coming from the sky. There's our cat, perched like a crow in the top of a large spruce. She's obviously distraught and can't come down.

The spruce is large and easy to climb. I climb up between the branches, and fifty feet up, there she is, overlooking the valley. When she sees me she purrs, jumps, and grabs on to my neck. As we descend, I can feel her sharp claws in my neck, like curved fir needles going straight into the skin.

I OFTEN THINK about what I want to teach my daughters. What's the most important thing I can give them in addition to being present and love?

Myself, I've lived my life in two worlds. One out in the woods, with everything that entails, and the other with fridge, freezer, and warm floors.

Our first few years together, Frida and I lived simply, my way. We met when I was down in Gothenburg for a while, and then she followed me up to Jämtland, where we bought a cottage that I then renovated myself. Frida had just finished her studies at the Design School in Gothenburg. Soon Signe was born, and Frida stayed home with her. Then came Helga.

We managed to get by on parental benefits and the hours I worked in health care. I always felt anxious when I was working at the old-age home, and I cycled or ran all the way to Järpen. Once home again, I'd spread that bad energy. The only thing I wanted to do was run hard intervals on the wet marshes beyond the house. I constantly ached for the next run. Earning a living for the family was a detraction from what I wanted to do. Frida was sad that I couldn't get my act together even for their sake.

She washed the children's clothes in a bucket in the yard or in a tub in front of the stove if it was cold. She fetched the water from the stream fifty yards from the house, a fresh water without the taste of soil, which she heated over the fire before washing with hand detergent. Her hands were dry and wrinkled, and she struggled everyday to keep it together.

We were so short of money that I sometimes looked for loose change on the floor of the car to buy oatmeal. When we couldn't afford gas, I ran to the store and hitched a ride home.

For me, I didn't mind going hungry or having it be 41 degrees inside the house—after all, it was still above zero. I thought Frida was soft and pampered when she was upset we couldn't buy more than food for the kids. It was as if my four years alone in the forest had eroded my understanding of what a family needs. That a family needs consideration and a structure that goes beyond temperature, seasons, and the black-and-white perspective nature provides. That life is about more than the most basic things.

The children got what they needed; they never lacked for anything. Frida is a good mom, and she created moments for the little ones they'll always remember as their childhood. But she herself had to sacrifice to give them what she could.

You can live a full life in simplicity, but you need to agree on exactly what that level of simplicity will be. The way we lived was not what she had imagined the simple life would be like.

I had a constant feeling of being a disappointment when I looked into her sad, pale blue eyes. In them was a mixture of love and incomprehension. Why is he doing this? What's wrong with him? Why can't he understand?

I had a family but hadn't checked into reality. I was only present mentally when I felt like it.

When I think back on that time, I am ashamed.

IT TOOK ME a few years to readjust to the modern world and to understand that life isn't just survival. Living can be about having a good life.

Today, Frida and I have a livelihood that has enabled us to both get a house on Öckerö and expand the cottage in Jämtland in stages. The family doesn't lack for anything. Today, I love our washing machine, not to mention the dishwasher. Of course they're unnecessary, and washing by hand is just as good, but the trade-off is less time to play with and kiss your children.

Although I no longer long for my forest life, I want people to see that there is value in both worlds.

It's nature that has shaped us into people. If we distance ourselves too much from nature, we won't understand why we need to care for it. It's difficult to value something that you have no relation to.

Of the approximately 350,000 years we've existed as a species, the past thirty-five years, or 0.01 percent of our existence, are about to make the planet unrecognizable due to our lifestyles.

But a human's time on earth is so short that we miss how quickly things change. We fool ourselves that more is always better than less, as if the entire point of existence is about maximizing everything.

When I went from running 10 kilometers in 33 minutes to 30 minutes, I doubled my workout load, from 100 kilometers per week to over 200 kilometers per week, an increase of 100 percent to get a 10 percent improvement. But I felt no greater sense of happiness than when I ran 10 kilometers in 30 minutes; I just wanted to run 10 kilometers in 28 minutes.

We talk about this kind of drive as if it were a good thing. That you should never be satisfied and that there's always something to improve. We'll squeeze the last 10 percent out of everything, even if the cost is 100 percent more. This is what motivates the people who run companies or work in politics and technology. Those who set the standard in society are the ones who want to be heard and seen. It's all about selling products and the next quarterly report.

Our rooster is called Grandma and is always angry. Sometimes, when our youngest daughter, Birgitta, collects the eggs, he sees his chance: he flaps air under his wings, scaring her into a corner, and jumps up to peck her. When she runs away crying, he chases after her, beating his wings at the thief. I think he'll soon be heading to a heavenly farm.

This wouldn't be a problem if the earth's resources were endless, but unfortunately, that's not how it works.

We put our trust in technology for solutions that enable us to continue the same way we live now. If only there were more efficient ways of consuming and inventions that turned fossil fuel energy into green, then everything will be alright.

But if you run too hard uphill, you hit a wall of lactic acid and have to stand at the side of the road just trying to breathe. Growing tomatoes in the same soil without adding nutrition leads to zero growth. If the energy output is too high, the technology you use simply doesn't matter.

Most farmers understand that the land sometimes has to lie fallow in order to recover. They have known it for ten thousand years. Maybe we should listen to them more? Listen more to those who live near things that grow, who understand that fall and winter are just as important as spring and summer?

We should live a little more like we did during 99.99 percent of our existence and appreciate what we already have. We should learn to discover everything available right outside the front door. You live where you live for a reason. Build a relationship with nature in your immediate surroundings. Invest time in that relationship. Use your body to feel.

The solution is the realization that more is not always better than less. The solution is to remember last summer, when the sun was shining and you biked to the lake and dove in and swam under the surface and felt the pressure from the water like a big hug.

You don't have to be afraid of tiring of it. Your experiences in nature will grow like the love for your children; it doesn't end just because you get more.

FIFTY YEARS AGO, Stephen Hawking said that we had one thousand years to find an alternative planet to live on to ensure our survival. Before he died, he adjusted that calculation to one hundred years.

He was one of the smartest people in the world, but I think it may be easier to solve our problems here on earth than to create a brand-new existence on another planet.

Artificial intelligence is now said to be the next step in development. Humanlike algorithms and robots will do what we don't want to do. A robot never skips work to care for sick children, is never ill, and can be repaired in half an hour if something breaks.

One way to predict the future is to draw a line from our origin to where we are now and see where it leads. It's called "piloting," and fishermen have used the method for thousands of years. You sail out from port and take a compass direction, then use two reference points to reach a third. So, where do we end up if we continue in the same direction?

If, today, 1 percent owns 90 percent of the earth's assets, then in the future, it will be 0.1 percent that owns 99 percent of the earth's assets. Capitalism and efficiency are the port and the compass direction. Environmental destruction and polarization are maybe not the goal, but the result.

We are almost nine billion people on earth. Do we really need another TV show to watch or a new kind of shoe insert for weak feet? A new planet?

In reality, we're probably just scared of death. We drive away thoughts of it by overloading our minds with more and more advanced devices and experiences. If our lives don't feel at risk, we don't need to think about the price of living the way we do.

FOG

When I was around twelve years old, I got my first boat, an orange Crescent with a black Mercury engine. It could reach twenty-two knots, and my friends and I often went out joyriding in our boats.

One spring day, when the sun began to warm up the sea, I wanted to go to the harbor at Björkö. The fog was thick and it was a stupid idea to go out, but it was just a few kilometers across the sound and I knew which way I was going, so I motored out of the harbor.

I had no compass, and in the fog no landmarks could be made out. It's easy to get lost in such conditions. I thought I was holding a straight course, but just drove around and around in one big circle.

When Grandpa and his brothers once went to Iceland to fish, they saw neither the sun nor the stars for seven whole days. Despite the conditions, they managed with the help of a compass and plumb line. That's what it takes when the fog is thick.

For me, the forest is a reference point that I can always trust. It helps me find my way when my mind is overrun by thoughts and emotions and when I don't know which direction to take.

But every day at midnight, the day ends. We're unable to save time for the future. Life is now.

Do we have more time left today than 35 years ago? Are we more present, do we have more meaning?

The faster you accept that death is as natural as birth, the faster you can start living. See the beauty of what is right in front of you and that which already exists within us.

THE KIDS HAVE a sports holiday, and Signe and I put on our skis. We've packed the sled with a saucepan and food for two days, sleeping mats and sleeping bags, ax, shovel, and saw. I pull the sled and Signe carries a lightweight backpack with hot water and some sandwiches.

We ski up to the heights around Lake Norsjön. It's minus 10 degrees and surface conditions are good with fifteen inches of powdery snow on a hard bottom. We easily move between the fir trees, while the sled floats on top of the snow.

Two hours later, we arrive at a small clearing with a view over Norsjön. A friend saw three bears here last summer, but now they're fast asleep in their dens. It's amazing how they survive for months without sleeping bags. They must be very rested, stiff, and hungry when they wake up.

I tie a tarp between two fir trees as protection against wind. Signe helps me trample the snow of the two-by-two-yard square that will be our base. While she shovels a pit for the fire, I fetch spruce branches to lay out like a thick bed. The branches are soft and warm to sleep on, with the scent of resin and forest that goes straight into your lungs.

When the bed is made and the wood collected, I light the fire. We melt snow into water and pour in red lentils, spices, and spicy sausage. We eat side by side while darkness falls. I was thirty when Signe was born. She's taught me a lot, pointed her index finger at me and said in a loud voice, "Earth calling Dad," when I've drifted away in my own thoughts.

We crawl into the sleeping bags and continue tending the fire until our eyelids are heavy and we can no longer keep it alive. As the fire dies down, I can make out the contours of the trees around us. Signe sleeps in a hat and breathes softly, lying on her side, just like her mother.

I think to myself that this is what I want to teach my daughters. That the forest and seasons can be a gateway to what's really important. This is where we humans come from, and it's important not to forget it.

I want to fill them with memories that reach far into their bodies, memories that they can return to anytime and anywhere.

Like last summer, when the eldest girls and I ran through the woods all the way down to the Indal River. The air was still, the sun was shining in a clear sky, and the spruce were dry, quiet, and thirsty. We jumped in below the Priest Falls and floated downriver towards Åkroken, where there's a good outcropping to get up on. The river was strong and warm, and we floated at a speed of two knots, Signe and Helga happy in the swirling water.

It didn't cost anything. It was just opening the door and going outdoors. A memory for life just a couple of miles from our house. Available to everyone.

A FEW MONTHS before Signe was born, I built a sauna. When she and Helga were small and the cabin lacked electricity, water, and bathrooms, we used the sauna several times a week to wash ourselves. Signe was one month old when she had her first sauna; we bathed her in a tub on the floor, then we laid her down on the lowest planks and let the warm air dry her.

Nowadays, we use the sauna as a storage room for skis and boots and as a sleeping place. I've put together two planks for a wide bed base and laid out a cold foam mattress on top; on the floor, a second mattress with a reindeer skin on top. In the winter, when one of the girls is bothered by something and having trouble sleeping, we walk through the snow in our underwear and hat and a sleeping bag under the arm. Then we lie down in the cold sauna with a couple of feet between us, close but still in our own bags. We can't hear each other's breath, only see the smoke from our mouths rising to the ceiling.

Out there in the darkness and silence, there's only the sound of our own hearts beating. That contrast between the heat in the sleeping bag and the cold outside reminds us of the way things used to be and how we live now.

I think that reminder is good for us humans. Life isn't always a smooth and fast highway; it's a forest path with roots and stones, a path that climbs hills and travels down over marshes and streams.

In the summers, we have an outdoor bed in the woods where we can lie, though more often we sleep under the open sky on the deck. When Frida goes to bed and turns off all the lamps in the house, the night sky and all its stars appear, like a thick layer of bright dots that becomes clearer the longer you look.

WASHING OUTDOORS

The feeling of standing naked in intense cold with a bucket of hot water at your feet, surrounded by snow-covered spruce and starry skies . . . Or in early summer, when the grass is high and the sky a dark blue with thunderstorms and lightning bolts zigzagging over the hills. What a bathroom!

Although we have a bath and shower indoors, I make sure to wash myself outdoors now and then. I always do it the same way: I fill a three-gallon bucket with hot water, bring along a bar of soap and a scoop, and head outside. If it's below freezing, I stand on a piece of wood, otherwise directly on the ground.

I squat and scoop hot water over my head and body. It's most beautiful in winter, when the contrast between inside and out feels most intense. I soap up, rinse away the foam, and head inside when the water runs out.

I stand naked in front of the stove, close enough that my skin feels too warm. I never use a towel, just let the radiating heat do the work.

I have a star for every death I've mourned. Mother's star is a thumb's width to the right of the easternmost visible star of the Big Dipper. She would have been sixty-one today and had twenty-two grandchildren. She got to meet five of them and has been gone for fifteen years.

The worry that I'll also develop MS like her is one I haven't known for a long time. I don't want to experience that panic again. But I've long since passed the age when her illness broke out, and now I know that I need to care for my own mind. I can't push myself too hard. I need to have a buffer of energy and power to handle that which hurts.

Grandpa's star is two fingers above the Polaris and visible only in winter when the sky is dark. When Grandpa, his face jaundiced with cancer, left this life, he felt at peace. There was no anxiety, just a few last calm deep breaths and a certainty that life would continue in another form.

He never got to meet my children. I can see his blue eyes in my brother's boy.

ONE EVENING I READ a book about horses for my daughter Helga before she went to sleep. She loves horses and wants to know as much as she can about different breeds, where they come from, and what they're used for.

Helga is small and thin and likes sweet things just like me. She loves to be in her body. Her psyche is like a birch branch in the spring when it blows: if pressured, she yields but doesn't break. She is gentle and wishes everyone well, but the big thoughts come creeping when night falls. She's always had trouble falling asleep. Just like me, it's easier for her to find rest in the fresh air.

We're lying in sleeping bags on the deck under the roof eave, Helga with an extra blanket over her. Signe is on a daybed beside her, reading a book, on a journey into her imagination.

All day, a cold fog filled Åredalen, and everything is covered in a thin layer of ice. My hands feel a bit stiff while I read to Helga about hot-blooded horses and cold-blooded horses and what sets them apart.

All apparently originate from the same horse, but have been influenced by the different environments they lived in. Hot-blooded horses come from warm areas and are lighter, faster, and more aggressive than cold-blooded ones, which lived in colder areas and are larger, slower, and calmer.

I think it's the same with us humans. Some people are watchers, less explosive, and take more time to react. Those qualities are just as important. I think we need more of that temperament for the times we're living in now.

ONE LATE WINTER evening when I run up Romo Hill, my chest feels heavy. My heart feels unwell, and there's no power in its beating. My legs feel slow and my breathing erratic. The thought occurs to me that I'm about to have a heart attack. Has my time come already? Can't I have just a few more years?

I've strained my body hard over the years, and the blood vessels are doubtless clogged with too much oatmeal, honey, and sugar. I see before my mind's eye how I fall down between the tree trunks, and I am filled with a strange sense of relief. This time, I'll probably stay in the woods for good.

I think of Frida and the children. They'll be alright. If they sell the house on Öckerö, they'll have food on the table for many years, and there's money in the wood on the property. I'm happy I crawled out from the ego well I was living in when the older kids were small, and that I eventually became a present dad and husband.

When I don't come home, Frida will call our neighbor Stefan. His dog will sniff some of my old underpants and be set loose. They'll find me in the frozen moss above the Romo marsh, a pleasant final resting place overlooking the Åreskutan. My head lamp will still be on but shining weakly. What was me is gone.

The thoughts surprise me, but they don't make me sad. If you feel at peace with dying, you can live life to the fullest.

It's then that my heart regains its rhythm. Energy returns to my legs. I run on.

)⁄

CLIMBING TREES

As a child, I climbed trees to pick eggs from bird nests and experience the feeling of standing on thin, tough branches high up in the tree canopy. Heights trigger and expand the imagination. You're aware that the branches could snap, and at any moment you could fall to the ground. Fear makes you feel more alive.

Now that I've passed forty, I'm still climbing trees. I do it for the exercise, for the view, to feel the branches of the tree yield under my weight, to watch the wind touch the leaves. On the ground, trees are experienced in a completely different way than when you're high up in one of them. Only then do you realize how huge they are. Your perspective changes.

To climb trees is to build a relationship. When the spruce needles have pricked your fingers and you've scratched your stomach against its rough bark, when you've felt the strength of the birch and breathed the scent of its leaves, then it's easier to understand that we need to respect the trees.

They're the ones giving us oxygen to breath.

—

For me, tree climbing serves as a yardstick for my physical and mental status.

Pines are incredibly tough to climb, and I almost never do. The

bark is rough but still slippery, and the trunk lacks strong branches all the way up to the crown. Pines are the homes of forest birds and eagles. They want to be left in peace.

Spruces have many branches that are easy to climb. The challenge with a spruce is agility and endurance, as well as flexibility as you twist your body around the branches when making your way up. Your hands will be pleasantly sticky from all the resin, and the needles will remind you just how many nerve endings are in your skin.

—

Deciduous trees require more strength and endurance since their branches are more spread out and there is quite a distance of branchless trunk between the ground and the canopy. Embracing the tree and pulling yourself up requires focus and pants that don't easily rip. In the Slottsskogen forest in Gothenburg, there are challenging deciduous trees with trunks as thick as car tires without anything to cling to. They're terribly hard, but really fun to climb. It's a real full-body workout, and your stomach, your arms, and the insides of your thighs will ache from the strain.

Some deciduous trees have very slippery bark; without friction it's up to just you and your muscles to counteract gravity, and it's tricky to find a place to rest. But then you reach the bottom branch, and everything immediately becomes easier. Always grasp the branches as close to the trunk as possible, where they're strongest, and you can pull yourself up using even the smaller branches without snapping them.

Sometimes when I feel weak and need strength training, I slip slowly down the trunk as soon as I reach the first branch. Then I start over. I do this for as long as I can.

——

Sometimes when I'm feeling stressed and deep in my own head, I find a young birch whose trunk is about as thick as my thigh. Thirty feet up, the trunk is no thicker than my forearms and sways under my weight, but a healthy birch never snaps. It's incredibly tough and flexible. I climb all the way up and stand on the last branch. I wrap my hands around the trunk, like I'm taking its pulse.

I stay like that until my mind has calmed and I no longer need its help.

TAPPING SAP

In the spring, when the birch begins to bud, it's time to drain a few liters of water from them. This doesn't damage the tree and leaves only a small wound that soon heals with a fine scar. For the birch, it's like a nosebleed.

—

You need a drill, a small tube, a string, and a bottle of some kind. Drill a hole as big as your little finger in the trunk, at an upwards angle and about three-quarters of an inch deep. Insert the tube into the bottle that you tie against the tree. A few hours later, it will be full.

The sweet birch sap contains nourishing amino acids, enzymes, and micro nutrients such as calcium, potassium, magnesium, and manganese. They are pulled from the earth when the sun and buds make the tree thirsty, and water begins to flow up through the trunk and out towards the branches, like a small stream defying gravity. After a few weeks, when the leaves are out, the rush is over and so is the sap.

In the springtime during my years in the hut, I used birches like a well. Not because I had to, but as a morning routine. To walk in the last snow still left between the firs and feel the warm, bright rays of the spring sun. To start the day with a mug of fresh, cold sap with the faint taste of trees and soil.

EDIBLE PLANTS

Despite my years in the forest, I'm certainly no expert on edible plants. I pick berries and the most common edible mushrooms, but when it comes to green things, I have only a few favorites that arrive when the snow has just disappeared.

1. Lady's mantle. I pick a fistful that I boil in water for a few minutes. It makes a tea that tastes like grass, but it's good with a little honey. You can also let it cool and use as a cold drink.

2. Birch leaves. When the leaves are small, I pick and eat them until my tongue turns green. They taste a little strong, but really sit well with my stomach.

3. Nettles. I fill a pot with nettles and then pour in water—barely half the pot—then gently boil a little. When the nettles start to fall apart, it's time to season with salt and white pepper. This soup can be eaten as is, but if you have a hand mixer, use it.

4. Spruce shoots. I eat them as they are. They taste like the forest and oil and have a strange aftertaste that lingers for a long time on the palate. Frida makes a syrup out of them by adding water, sugar, and lemon and then leaving them to soak for a few days, like you would elderflower syrup. Very refreshing.

DOING YOUR BUSINESS OUTDOORS

During the years I lived outdoors, I had the world's best toilet: grand views, fresh air, peace of mind, and constant change. Toilet paper is just a waste of resources that 80 percent of the world's population does without. You use what nature has to offer, or the left hand and water, and wash thoroughly afterwards.

Practice makes perfect, and for the last twenty years I've done most of my business outdoors. Here's some advice from my experience bank.

—

Autumn is the absolute best time to do your business outdoors. There's a lot of soft materials with which to wipe: withered leaves, green moss, and peat moss. The peat moss is like an antiseptic washing sponge, not to mention soft against your butt. Try to get as far from the nearest fir as possible: spruce needles are prickly and they have a habit of finding their way inside moss. Nobody wants the surprise of a stinging needle in their behind. Just pull down your pants and get started. However, you'll find that a brook can serve as a giant bidet out in the woods. It's fantastic! The disadvantage is that relieving and washing yourself in running water may spread bacteria downstream, where someone might be refilling their water bottle. However, if you're far out in the middle of nowhere and know that no one is nearby, your best bet is the creek.

The late fall, before the snow has come, is the worst time for going outdoors. Everything is rock solid and stuck in the ground. You can give spruce cones a try, but pinecones are scratchier, if you ask me.

Then comes the first snow, and it's brilliant. The snow is wet and you don't need anything else.

This is followed by the really cold periods with powder snow. So, I have an observation about this: it seems like one's behind isn't very sensitive to cold temperatures. I don't mean the cheeks, but the area where the sun never shines. It can handle most temperatures. Drying yourself under these conditions would be a breeze if it weren't for frozen hands. When it's minus 31 degrees, it's easy to get a little frostbitten, so it's important to be quick about it.

After that, when there are warm days and freezing nights, that's when a thin ice forms over the snow. It's difficult to dry yourself with something rock hard and glossy. You will have to punch a hole through the ice with your foot and hope there is a softer layer of snow underneath.

—

Then comes spring and summer, a wonderful time for doing your business outdoors. The streams start flowing again, the moss makes a comeback, the trees bloom, and green beautiful grass begins to germinate. These seasons give your butt time to recover. But there are certain types of grass you should avoid: round meadow grasses that are difficult to clean with, and that broad, flat grass that is like razor blades and leaves small cuts on your behind. Go for the short and soft grasses. Ferns also work well.

DREAMY ON YOUR BACKSIDE

Peat moss has slightly different coloring depending on where it grows, from yellow to green. The color doesn't matter; it does its job regardless. Many people think that white moss is what we use in Christmas ornaments, but that is actually lichen, which won't wipe very well.

SLEEPING OUTDOORS IN THE CITY

When I'm down in Stockholm for work, I choose between a few different sleeping places depending on what I want to do. During a quick
visit, I sleep in the Lill-Jansskogen forest or northern Djurgården,
close to the city but still among trees. If I want to stay a little longer,
I go out to Lidingö or down to the area around Hellasgården, where
there isn't a lot of traffic noise.

———

As I walk through the city on my way to one of my sleeping places,
I often meet my less fortunate brothers, the homeless. It's a strange
feeling to see them lying under blankets under lit windows on busy
streets.

For me, it's a privileged, voluntary choice to sleep outdoors. In
the past, the forest and the loneliness were a personal choice that
allowed me to shed everything until only the most basic remained.
Without that journey, I wouldn't have calmed my mind. The homeless
struggle with that same cold and basic needs, but probably experience the outdoors in an utterly different way. For them, the sidewalk
was all that was left when the door into the warmth was closed.

Dressed in tattered clothes and with tired eyes, they struggle
with abuse, grief, and failing relationships. I've often thought that
I myself could easily have ended up in similar circumstances, that

sometimes it takes so very little for a person to fall, and land hard.

Sometimes, I have the naïve impulse to ask if they want to join me for a couple of miles, go out in the woods where they might feel more relaxed under a spruce and sleep on soft moss instead of on asphalt. It wouldn't begin to address the institutional and social causes of their homelessness. But at least there'd be birdsong instead of traffic noise.

—

It took me a couple of years to find my favorite places. Now they're so familiar that even blindfolded I'd recognize them by smell and sound alone. It fills me with a sense of security to return to a place where I know I'll find deep sleep.

In addition to clothes, I carry a headlamp, sleeping bag, sleeping mat, and a tarpaulin that I can tie up in case of rain or snow. Water, fruit, and some nuts are sufficient provisions.

In the Lill-Jansskogen forest, I never make a fire. Rather, I try to blend in like an animal. The closer to the city I am, the less light I create—that's my rule. In an urban forest, I hardly use my headlamp. I don't want anyone to know where I am. I feel safest when it's just me, the dark, and quiet trees.

—

If you want to sleep outdoors for one night in Sweden, there are forests at walking or cycling distance from almost every urban environment. You don't even have to leave the larger cities. In Gothenburg, Änggårdsbergen is just a stone's throw from Sahlgrenska and the Botanical Garden.

Plan to leave home so that you arrive no later than an hour before sunset—it may take some time to find a place that works. Go deep enough into the woods where you can't see lights or houses anymore. Ideally, you shouldn't hear any traffic.

During the bright and warm months, I look for places with a view so I can see far into the distance. In the winter, I seek out flat surfaces surrounded by dense spruce forest; the trees protect against the wind and rain. The colder it gets, the more protection you'll need.

—

Once you find a sleeping place, gather some branches and make a fire. Heat tea water in a pot over the fire, and grill a sausage. When darkness comes, pull out your sleeping mat and sleeping bag and crawl in.

It can feel a little uneasy as the fire burns out and you're surrounded by darkness and everything becomes quiet. But even if it's tough to be in the dark, there's nothing to worry about.

Moose, deer, and lynx know where you are, but stay away. They don't want to harm you. And people?

I've slept outside in Stockholm, Gothenburg, Norrköping, Örebro, Umeå, Mora, Hallsberg . . . Once I've laid down, I've never encountered a single person. Who would it be? No one else is in the forest at night.

There's only one kind of creature you need to watch out for, and only on the east coast of Sweden: ticks.

THE OUTDOOR BED

You can build an outdoor bed anywhere—in a garden, in a forest glade. The most important thing is that the place is accessible and that it's not difficult to walk there. If you decide to sleep outdoors, you should be able to do so without having to plan two hours in advance.

—

I built our family's outdoor bed a bit away from our house. We need just walk down the hill, across a road, and then for a minute along an unpaved trail through the forest, and then we're there.

I built it for two people. The main end where you rest your head is to the west, so that the evening sun shines on your face. A few meters away, there's a small ring with stones for a fire.

We go there year-round, especially if any of the girls need extra quality time. It's easier to talk about certain things when lying down next to each other in sleeping bags, the heat of the fire or the evening sun on the children's faces.

You're close to each other, but there's also room for silence.

COLLECT

If you want to build an outdoor bed for two measuring 55 x 83 inches, you will need:

- 6 x 32-in. poles (bed legs)
- 2 x 90-in. and 2 x 63-in. poles (bed frame)
- 30 x 63-in. poles (bed base)
- A sturdy cross bar at the top (87 in.), as well as 2 x 70-in. poles that you'll insert in the ground (for the roof construction)

The poles can be made from trees or wood lying about. Not half rotten, not too knobby (if they are, use your ax to straighten them).

THE FRAME AND THE BASE

- Sharpen the bed legs at one end.

- Hammer them into the ground, a pole in each corner and two more for the middle of the long sides.

- Tie the poles to the long side of the bed (the poles should be about 8 in. longer than the bed so that it's easy to tie everything in all directions).

- Do the same with the short side; attach one pole above and one below the long side lines. If you don't have an eye for it, bring a level so you can be sure the bed is balanced.

- Then place the poles in the bed base, leaving about the same width of space between the poles as they are thick.

FASTENING THE BASE

This can be done several different ways, but it's not really more difficult than pulling a few turns of rope around what needs to be tied together. Keep the distances neat and even, and pull the rope tight. Avoid protruding branch remnants and heavy growth. Use an ax to trim the poles.

MAKING THE CANOPY

If you haven't already, stand on the bed and hammer the long, sharpened poles for the roof into the ground. Tie to the frame. Tie the cross beam at the top, but first check that it's strong and even—otherwise it won't hold up the tarp or can cause holes in it.

THE MATTRESS

Start with larger spruce branches and build up with thinner. That way you'll have a nice, soft bed.

THE ROOF

Fabric or tarpaulin, it makes no difference. Just use whatever you have. However, fabric may need to be waxed to withstand rain. Hammer four stakes into the ground a bit away from every bed corner. Throw the cloth over the cross beam at the top and spread out to the corners.

MOBILE OPTIONS

I also have a folding daybed that I keep with me if I travel around the country in the car giving lectures. It's a quick and easy option for overnight or a rest anywhere.

DRYING FIREWOOD

In the springtime, people cut down trees and chop them into firewood. This allows the wood to dry during the summer and autumn, and it's ready to use when winter comes.

Myself, I don't like to spend time making firewood in the spring, when the weather and skiing are at their best in Jämtland. It's the "fifth" season: cold at night and still a lot of snow in the mountains, but warm and usually mild during the day. With the spring sun on your back, you can go for long walks with only a thin wool sweater.

Instead, I prefer to work with the wood when the grass is high and the days are long. I fell birch trees when the leaves are at their largest, and then leave them for a while. The leaves will suck the liquid from the tree and then shrink and fade as the trunk becomes dry.

—

Birch and other deciduous trees have higher density and hold more energy than conifers. You can easily feel it yourself: hold a birch log in one hand and pine in the other. The birch is definitely heavier. Therefore, it burns longer and throws off fewer sparks. Pine and spruce are more common and quick to roar into a bonfire, but they don't burn for very long.

No matter what kind of wood you choose and what season you chop the wood, stack the logs this way and that way on a wooden pallet or something similar that keeps them a bit off the ground. If you make sure the top layer of logs are bark side up, it's not necessary to cover the pile. That said, it never hurts to protect the pile with a roofing sheet. The important thing is that you don't cover the sides, because the sun and the wind still need to be able to do their jobs.

—

When I lived in the hut, I had no house to heat and didn't need a lot of wood, just a few cubic yards a year. If the wood began to run out, I just went out into the woods and found a dry spruce. Preferably, a wilted tree that had been dead for many years, but with such strong roots that it was still standing upright. To test it, I hit it a few times with the back of the ax against the trunk, and if the tree answered with a high pitch, the tree was dry. Ready to fell and chop right away.

THE STUMP

To sit on a stump and stay there until the restlessness and fight-or-flight response disappear, till you begin to feel that you're good just the way you are, that is to stump.

You can stump anywhere and on anything—the stump is just a symbol of the idea that you sometimes have to leave your head and reach down into your heart. But even if you can stump on a chair or stool, it's my honest belief that the effect will be greater if you have a real stump.

Because it may not work to take time off from work or family to go into the forest and just be, let the stump come to you. Just go out into the countryside and find a big, red building—that's where they live, the farmers who provide us with food. It will be a bit of bad luck if you manage to find a farmer who doesn't have a stump to spare. Give the farmer ten or twenty dollars, and you'll get a stump to take home with you.

Make sure it's no higher than fourteen inches, because then it will feel like a chair. A chair is easy to get up from. You should sink down on a stump. If it has a diameter of twelve to sixteen inches, it will fit your behind perfectly.

—

A classic spruce stump is dry and quite hard and quickly loses its scent. A discreet and safe choice. Pine stumps are a little softer and give off a soft and oily forest smell for months. The disadvantage is that the resin readily sticks to your pants like chewing gum. The birch stump emits less of a scent but lacks resin and becomes very beautiful when the white, shimmery bark has dried for a few months.

———

Keep the stump in your living room or at work. Sit on it when you have too much in your head. Insist that whoever is sitting on the stump mustn't be disturbed.

Behind our house in Jämtland, there are plenty of windfalls I can cut into stumps. When I'm out lecturing or greeting people, I often take one with me and give it as a gift.

I think the stump can help us access what lies beyond what we can see.

THANKS

To Signe and Helga, for being such wonderful big sisters
when we were out taking pictures late in the evening!
And to Birgitta, of course!

To the older tribe, you who have gone before us
and laid the foundation for what we have now.

To all the friends who are dear to us, it's good to
have you by our side. If only everyone could have
such wonderful friends.

To Offside Press, you are unbeatable.

MARKUS TORGEBY was a promising, elite runner until an injury ended his career and triggered a life crisis. Four years of self-selected loneliness in a hut in the Jämtland forest returned a calm to his body and gave him a new sense of direction. Today, Markus lectures on "what is really important" and builds houses and outdoor beds. His autobiography, *The Runner* (2015), has sold almost fifty thousand copies in Sweden and has been translated into half a dozen languages.

FRIDA TORGEBY created the photographs and illustrations in *Under the Open Skies*. She has studied at the Photography College and has a degree from HDK, the College of Design and Crafts, in Gothenburg. Since then, Frida has, among other things, worked as an expert senior designer at Volvo Cars, but now runs her own photo and design business. Frida and Markus also run the Torgeby clothing brand (Torgeby.com) together.

Markus and Frida have three daughters and live in Jämtland, outside Undersåker. You can follow and reach them via Markustorgeby.se and on Instagram.